PERENNIAL MATH

Tournament Tests 2

Grades 7-8

PERENNIAL MATH TOURNAMENT TESTS

These are a collection of tests from past Perennial Math Tournaments. Each tournament has two rounds of testing. Each student takes an Individual Test during the first round and then students combine with their teammates to take a Team Test during the second round. The solutions included in this book are only a short sample of the many possible solutions that could be used to solve each problem.

INDIVIDUAL TESTS GRADES 7-8

30 minute practice tests for individual students

Grade 7-8 Individual Event (30 Minutes)

1) Larry's average on 6 tests was 89. If he wants to make an average of 90, what must he score on the next test he takes?

2) In this magic square the sum of the numbers in each of row, column, and diagonal is the same. What number should be in the box marked X?

X		3	16
	15	10	
	1	8	
7	12		

3) A picture frame is 18 x 20 inches. There is a 2 inch mat around the picture. What is the area of the mat (shaded region)?

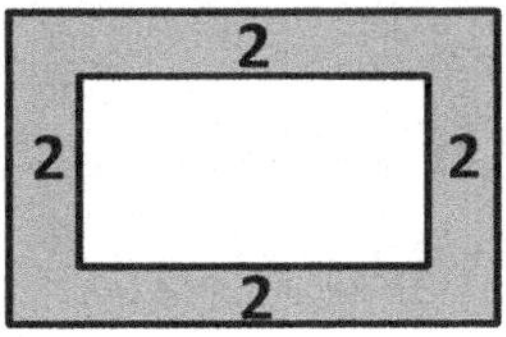

4) If a roller coaster can accommodate 75 people in 25 minutes, how many people could ride the roller coaster in 2 hours?

5) A set of 10 consecutive even numbers has a sum of 250. What is the third number in the set?

6) What number multiplied by itself is equal to the product of 12 and 108?

7) A certain number is divisible by three and also by eight. When the number is divided by 7, the remainder is 2. What is the smallest number between 200 and 300 that satisfies all these conditions?

8) Sally shipped a package that weighed *W* pounds, with *W* being a whole number. To ship this package it costs a total of $1.85 for the first 5 pounds and .19 cents for each additional pound. If the package cost $6.41 to ship, how many pounds did it weigh?

9) A dune buggy has a front wheel with a circumference of 4 ft. and a back wheel with a circumference of 5 ft. How many more turns will the front wheel make than the back wheel in 2 miles? *1 mile = 5,280 ft.*

10) The numbers given represent the sum of the objects in each row or column. Find the value of the middle row.

PERENNIAL MATH

11) Find the slope of the line segment joining the points (4, - 3) and (- 3, 4).

12) 6 + 8 x 2 – 4 x 3 =

13) Find the difference between the sum of the three highest and the sum of the three lowest prime numbers below 60.

14) Sixteen people shake each other's hand only once at a party. How many handshakes take place altogether?

15) Suppose five days after the day before yesterday is Friday. What day of the week will tomorrow be?

INDIVIDUAL ANSWER DOCUMENT (Grades 7-8)

SCORE

Team # ___________ School ____

Student Name____

Student Number__________________

1. ____________	6. ____________	11. ____________
2. ____________	7. ____________	12. ____________
3. ____________	8. ____________	13. ____________
4. ____________	9. ____________	14. ____________
5. ____________	10. ____________	15. ____________

Grade 7-8 Individual Event Solutions

1) Since he had an average of 89 on 6 tests his total points scored was 534. In order to make a 90 average on 7 tests, he must obtain a total of 630 points. Larry must score 630-534 = 96 points.

9	6	3	16
4	15	10	5
14	1	8	11
7	12	13	2

2) Adding up the 4 numbers in the diagonal the sum is 34. Now you can solve for the top box 34 = (12 + 1+ 15 + ?). The top box is 6. Now to solve for X subtract the 3 numbers from 34. 34 – (6 + 3 +16) = X
X = 9

3) The area of the entire frame is 18 x 20 = 360cm² and the inside area of the frame is 14 x 16 = 224cm². Subtract the inside area from the total area and that will give you the area of the surrounding mat. 360 – 224 = 136cm²

4) X:120 = 75:25. X/120 = 75/25. 25X = 9000 or X = 360. The roller coaster can accommodate 360 people in 2 hours.

5) The sum of 10 consecutive odd numbers is 250. Divide 250 by 10 and the average is 25 which would fall between the 5th and 6th numbers. So the 5th number is 24. The numbers in the set are 16, 18, 20, 22, 24, 26, 28, 30, 32, 34. The third number in the set is 20.

6) 12 x 108 = 1296. The square root of 1296 is 36. The number multiplied by itself is 36.

7) Since the number is a multiple of 3 and 8, then it is a multiple of 24. List these and numbers that are multiples of 7 adding 2 and find the smallest number between 200-300. The answer is 240.

Multiples of 3 and 8	216	**240**	264	288		
Multiples of 7+2	205	212	219	226	233	**240**

8) $6.41 - $1.85 = $4.56 The package weight for the remaining pounds can be calculated by 4.56/.19 = 29. So there were 5 pounds ($1.85) + the additional 24 pounds ($4.56) or 29 pounds.

9) Two miles would be 10,560 ft. Divide this by 4 and the front wheel turns 2,640 times. Divide this by 5 and the back wheel turns 2,112 times. To find out how many more times the front wheel turns than the back wheel subtract 2,640 – 2,112 = 528 times.

10) Use the first column to solve for the star. 3stars = 33 so 1 star=11. Substitute this in the first row to solve for the heart. 3 stars + 1 heart = 48. So 1 heart = 15. Solve for the X by substituting the value of a heart (15) in the second column. So 2X + 15 = 27. X is worth 6. Now to solve for the middle row: 1 star + 1 X + 2 hearts = 11 + 6 + 30 = 47.

Grade 7-8 Individual Event Solutions

11) $x_1 = 2, y_1 = -4, x_2 = -4$, and $y_2 = 2$. $\text{slope} = m = \dfrac{y_2 - y_1}{x_2 - x_1}$

$$\frac{4-(-3)}{4-(-3)} = \frac{7}{7} = \mathbf{1}$$

12) Use the order of operations PEMDAS.
6 + 8 x 2 – 4 x 3 =
6 + 16 – 12 = 10

13) (59 + 53 + 47) – (2 + 3 + 5) = 159 -10 =149

14) Since there are 16 people the answer is 15 + 14 + 13 + 12 + 11 + 10 + 9 + 8 + 7 + 6 + 5 + 4 + 3 + 2 + 1 = 120 handshakes.

15) Draw a diagram to help solve. Suppose five days after the day before yesterday is Friday. What day of the week will tomorrow be?
Tomorrow will be Wednesday.

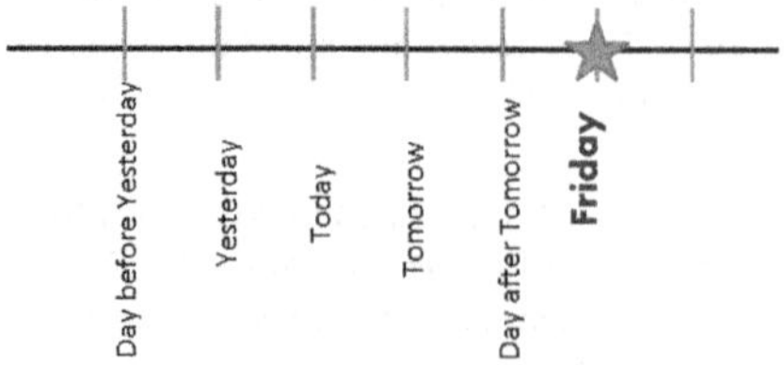

INDIVIDUAL ANSWER KEY

(Grades 7-8)

Instructions for Scoring:

- *Mark CORRECT answers by circling "O" the question number, and INCORRECT answers with an "X". Each correct answer is worth one point. Count the circles (correct answers) and write that number in the "SCORE" box.*

CORRECT → (1.) ________

INCORRECT → X. ________

1. 96	6. 35	11. 1
2. 9	7. 240	12. 10
3. 136	8. 29	13. 149
4. 360	9. 528	14. 129
5. 20	10. 47	15. Wednesday

PERENNIAL MATH

Grade 7-8 Individual Event (30 Minutes)

1)Three consecutive integers have a sum of 492. What is the product of the three integers?

2) The area of ΔAFG is twice the area of square DEFG. AF and AG meet DE at B and C respectively such that BC = 6cm. If the side length of the square is 8cm, what is the area of ΔABC?

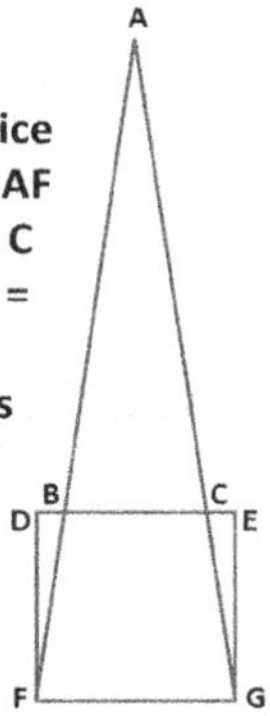

3) Beads are placed on the string in the following sequence: 1 red, 1 green, 1 blue, 2 red, 2 green, 2 blue, with the number of each color increasing by one every time a new group of beads is placed. How many of the first 160 beads are

4) Eric gave Craig and Trevor as many marbles as each had. Craig then gave Eric and Trevor as many marbles as each had. Then Trevor gave Eric and Craig as many marbles as each had and they all had 24. How many marbles did Eric have to begin with?

5) There are 12 candidates in a class election. The winner will be the president, and the runner-up will be the vice-president. How many different combinations of president and vice-president are possible?

6) Tristan is currently paid $480 for working a 50 hour week. His weekly salary is to be increased by 10% but his hours are to be reduced by 10%. How much less will he be making each week?

7) Marla's three children's ages are all different prime numbers. The sum of their ages is 41 and the difference between 2 of the ages is 16. How old is her oldest child?

8) The Ziefer farm has only chickens and pigs. When the farmer counted the heads of the stock in the farm, the number totaled up to 200. However, when the number of legs was counted, the number totaled up to 540. How many more chickens were there in the farm?

9) Each of the digits 1, 2, 3, 4 are used once to create a four-digit number. Twenty-four different four-digit numbers can be formed this way. If these twenty-four numbers are then listed from the smallest to the largest, in what position is the number 3214?

10) In a survey of 50 people, 4 people liked pizza and burgers but not spaghetti. 12 people liked spaghetti and burgers but not pizza. 10 people liked pizza and burgers. 18 people liked pizza and spaghetti. Altogether, 30 people liked pizza, 32 people liked spaghetti, and 28 people liked burgers. How many people like burgers only?

PERENNIAL MATH

11) Find the slope of a line passing through (0, -2) and (3, 4)

12) If x □ y is defined as $(x + y)^2 - 2xy$, then 5 □ 7 =

13) What is the units digit of 13^{32}?

14) What is the quotient when 16,000 is divided by $4^3 \times 5^2$?

15) What is the value of *n* in this equation: $2 \cdot 4 \cdot 8 \cdot 5 \cdot 5 \cdot 5 \cdot 5 \cdot 5 \cdot 5 = 10^n$

INDIVIDUAL ANSWER DOCUMENT

(Grades 7-8)

SCORE

Team # __________ School ______________________________

Student Name ______________________________

Student Number D - __________

1. ________	6. ________	11. ________
2. ________	7. ________	12. ________
3. ________	8. ________	13. ________
4. ________	9. ________	14. ________
5. ________	10. ________	15. ________

PERENNIAL MATH

Grade 7-8 Individual Event Solutions

1) Three consecutive integers have a sum of 492 so the average of those 3 integers is 164. The three integers are 163, 164, and 165. Their product is 163 x 164 x 165 = 4,410,780

2) Find the area of the square DEFG, the area of ΔAFG, and the area of trapezoid BFGC. The area of the square is length x width = 8x8 = $64cm^2$. The area of ΔAFG = 2 x square DEFG = 2 x $64cm^2$ = $128cm^2$. The area of the trapezoid, (BC + FG) x DF ÷ 2 = (6 + 8) x 8 ÷ 2 = 14 x 8 ÷ 2 = $56cm^2$. Now (subtract area ΔAFG – area of trapezoid BFGC) to find area of ΔABC. $128cm^2$ - $56cm^2$ = $72cm^2$. The area of ΔABC = $72cm^2$.

3) The greatest total that can be placed with equal numbers of red, green and blue beads is 3(1) + 3(2) + 3(3) + ... + 3(9) = 3+6+9+12+15+18+21+24+27 = 135. At this point there are 135 ÷ 3 = 45 beads of each color. We have counted up to a group of 9 beads of each color and 160 -135 = 25 beads left over. So there will be 10 red beads, 10 green beads, and the remaining 5 beads will be blue. There are 1 + 2 + 3 + 4 + 5 + 6 + 7 + 8 + 9 + 5 = 50 blue beads.

4) If they all ended with 24 marbles, then you know that the starting amount totals 72. Work backward to solve what each person had in the previous step. Eric had 39 marbles to start with.

	Eric	Craig	Trevor
They end with 24 each, but Trevor had given Eric and Craig as much as each had.	24	24	24
The result of Craig giving Eric and Trevor as much as they each had.	12	12	48
The result of Eric giving Craig and Trevor as much as they each had.	6	42	24
Starting amounts add up to 72.	39	21	12

5) There are 12 candidates in a class election. The winner will be the president, and the runner-up will be the vice-president. Once a candidate is picked then they cannot be picked for the other position. So to choose president, there are 12 choices, then once someone is selected, that leaves 11 people remaining for vice-president. So 12 x 11 = 132 possible different combinations of president and vice-president.

6) Calculate his current pay rate and subtract the new pay rate and new time. $480/50 = $9.60. His new pay is 10% more which is .96 + 9.60 = $10.56. However he will be working 10% less hours which is 50 x 10% = 5.0 hours. 50 – 5 = 45 hours x $10.56 = $475.20. He will be making $4.80 less.

7) Marla's three children's ages are all different prime numbers. The sum of their ages is 41 and the difference between 2 of the ages is 16. How old is her oldest child? List prime numbers and look for 2 of them that have a difference of 16. {2, 3, 5, 7, 11, 13, 17, 19, 23, 29, 31} 3 and 19, 7 and 23, 13 and 29 are all pair that have a difference of 16. 3 + 19 = 22, but 41 – 22 = 19. That doesn't work because the ages must be different prime numbers. 7 + 23 = 30. 41 – 30 = 11. The children are 7, 11, and 23. The oldest is 23.

8) Make a chart. There were 130 chickens and 70 pig. 130 – 70 = 60 more chickens than pigs.

Pigs	100	90	80	70
Chickens	100	110	120	130
Legs x 4	400	360	320	280
Legs x 2	200	220	240	260
Total heads (200)	200	200	200	200
Total legs (540)	600	580	560	540

PERENNIAL MATH

Grade 7-8 Individual Event Solutions

9) There are 24 four-digit numbers that can be made from the digits 1, 2, 3 and 4. There are 4 possibilities for the first digit so there are 24 ÷ 4 = 6 numbers that begin with a 1. The numbers are 1234; 1243; 1324; 1342; 1423 and 1432. So there are 6 numbers that begin with each digit. So list out the numbers that begin with 3 in order to find out what position it comes in. 3124, 3142, <u>3214</u>. Since there were 12 numbers already and now these 3, then 3214 is the 15th number.

10) All regions total 50 people. We are solving for Region II. Label the totals for each category:30 people liked pizza 32 liked spaghetti, and 28 liked burgers. Fill in these areas: 4 people liked pizza and burgers but not spaghetti. 12 people liked pizza and spaghetti but not burgers. Use these to solve for missing areas:10 people liked pizza and burgers [Region V + 4]. 18 people liked burgers and spaghetti [Region V + 12]. Once you solve for Region V then you can solve for the burgers to get a total of 28. 28 – (4 + 6 + 12) = 6 people who like burgers only.

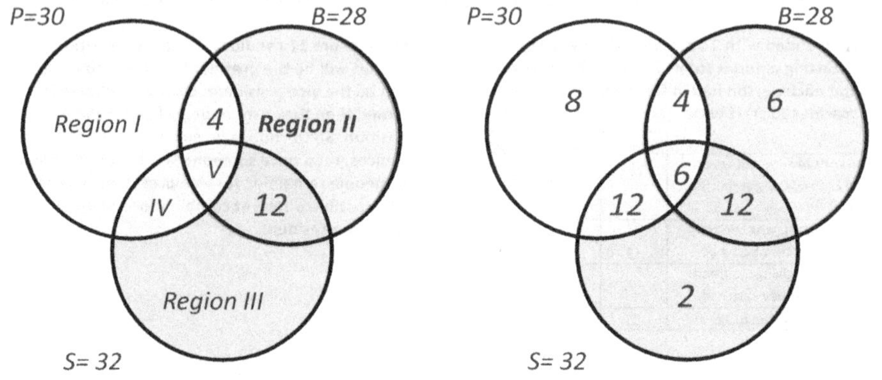

11) Find the slope of a line passing through (0, -2) and (3, 4)
$m = y_2\text{-}y_1/x_2\text{-}x_1$
4 – (-2)/3-0 = 6/3 = 2

12) If x □ y is defined as $(x + y)^2 - 2xy$, then 5 □ 7 =
$(5 + 7)^2 - 2(5)(7) = 12^2 - 70 = 74$

13) What is the units digit of 13^{32}?
$13^1 = 13$
$13^2 = 169$
$13^3 = 2197$
$13^4 = 28561$
$13^5 = 371293$ 13^{32} would end with a 1 in the units digit because the pattern (3, 9, 7, 1) repeats itself.

14) What is the quotient when 16,000 is divided by $4^3 \times 5^2$?
$4^3 \times 5^2 = 64 \times 25 = 1600$. 16000/1600 = 10

15) What is the value of *n* in this equation: $2 \cdot 4 \cdot 8 \cdot 5 \cdot 5 \cdot 5 \cdot 5 \cdot 5 \cdot 5 = 10^n$
$2 \cdot 4 \cdot 8 \cdot 5 \cdot 5 \cdot 5 \cdot 5 \cdot 5 \cdot 5 = 1{,}000{,}000$ which is 10^6 Therefore, $n = 6$

PERENNIAL MATH

INDIVIDUAL ANSWER KEY
(Grades 7-8)

Instructions for Scoring:

- ***Mark CORRECT answers by circling "O" the question number, and INCORRECT answers with an "X". Each correct answer is worth one point. Count the circles (correct answers) and write the number in the "SCORE" box.***

CORRECT → (1.) ________

INCORRECT → ~~1.~~ ________

1. 4,410,780	6. 4.80	11. 2
2. 72	7. 23	12. 74
3. 50	8. 60	13. 1
4. 39	9. 15	14. 10
5. 132	10. 6	15. 6

Grade 7-8 Individual Event (30 Minutes)

1) Two regular dice are rolled. What is the probability that the product of the two numbers on the tops is greater than 8? Answer as a reduced fraction.

2) If this pattern were to continue, how many cubes would be in the 6th shape?

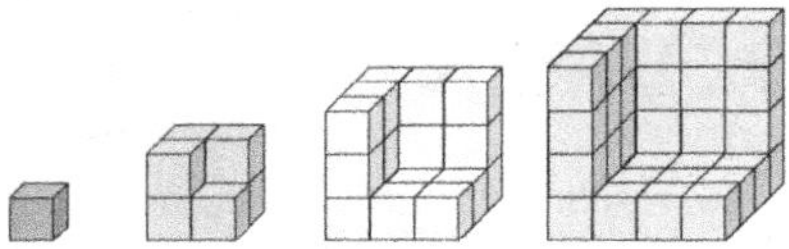

3) The cubes are glued together and dropped in a bucket of paint. How many cubes would have only 3 sides painted if the shape came unglued?

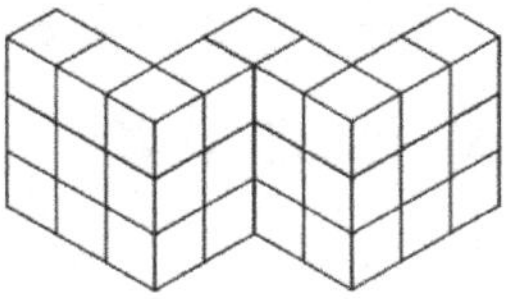

4) There are 1127 jellybeans in a bag. What is the least number of jellybeans that Tanya would have to remove from the bag, so that the remaining jellybeans can be divided equally among her and 7 friends?

5) Of all the numbers between 1 and 100, how many times does the digit "1" appear?

6) Ann and Betty have \$19 together. Betty and Cathy have \$27 together. Ann and Cathy have \$22 together. How much money does Cathy have all by herself?

7) Mary ate a total of 65 jellybeans in a number of consecutive days. Each day she ate 5 more jellybeans than the day before. How many jellybeans did she eat on the last day?

8) There are 12 students in the algebra class. On a recent test, the class average was 80. When Ann asked for her score, the teacher could not find it, but she did remember adding Ann's score in the average. She also remembered that Ann had passed. These are the scores of the other students: 65, 65, 70, 70, 80, 85, 85, 90, 90, 90, 100. What was Ann's score?

9) The units (ones) digit of the product of any six consecutive whole numbers is what?

10) The product of three consecutive integers is 210. What is the sum of the three integers?

PERENNIAL MATH

11) Find the perimeter of the right triangle with legs that have lengths $x + 2$ and $x + 6$. The hypotenuse has length $2x$.

12) $6 + 2 \times 7 - 8 \div 2 =$

13) Find the slope of the line segment joining the points (-2, - 1) and (4, 3). Reduce the fraction to its lowest terms.

14) Find the GCF of 189 and 621.

15) Express 5.556×10^6 in standard notation.

INDIVIDUAL ANSWER DOCUMENT

(Grades 7-8)

SCORE

Team # __________ School ____________________

Student Name ____________________

Student Number __________

1. __________	6. __________	11. __________
2. __________	7. __________	12. __________
3. __________	8. __________	13. __________
4. __________	9. __________	14. __________
5. __________	10. __________	15. __________

PERENNIAL MATH

Grade 7-8 Individual Event Solutions

1) Make a chart. The probability of the product of the two dice being greater than 8 is 20/36 = 5/9.

	1	2	3	4	5	6
1	1	2	3	4	5	6
2	2	4	6	8	10	12
3	3	6	9	12	15	18
4	4	8	12	16	20	24
5	5	10	15	20	25	30
6	6	12	18	24	30	36

2) The pattern is the number the shape is in the sequence cubed minus the number before it cubed. So the 6th shape would equal $6^3 - 5^3 = 216 - 125 = 91$ cubes

3) There would be 16 cubes with 3 sides painted.

Plus 1 block that is hidden

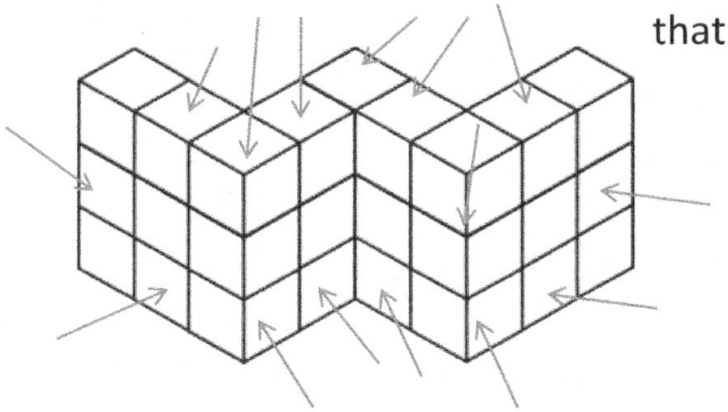

4) She would have to remove 7 jellybeans so that the jellybeans could be divided equally among her and 7 friends. 1120 ÷ 8 = 140 jellybeans each.

5) The digit 1 appears 19 times between 1-100. Exclude 1 and 100. 10, 11, 12, 13, 14, 15, 16, 17, 18, 19, 21, 31, 41, 51, 61, 71, 81, 91. Don't forget to count it twice in the number 11.

6) A + B = \$19. B + C = \$27. A + C = \$22. Add together 2A + 2B + 2C = \$68. Divide both sides by 2, then A + B + C = \$34. Substitute in the amounts for (A + B + C) – (A + B) = \$34 - \$19 = \$15. Cathy has \$15.

7) Make a chart. Start with her eating 1 jellybean the first day. She ate 23 jellybeans on the last day. Another possible solution is 30 jellybeans on Day 1 and 35 on Day 2.

Day 1	Day 2	Day 3	Day 4	Day 5	Sum
1	6	11	16	21	55
2	7	12	17	22	60
3	8	13	18	23	65

Grade 7-8 Individual Event Solutions

8) Average = sum of items/number of items. Twelve scores gave an average of 80 so the sum of all the scores must be 80 x 12 which is 960. The sum of the eleven scores the teacher had written down was 890. She had to score 960 – 890 = 70 points.

9) Regardless to what six consecutive numbers you use, the product will always end in zero because you will be multiplying by a factor of 5 and an even number.

10) Use factoring because the question asks for the product which is the opposite of factoring. Since the factors of (210) = 2, 3, 5, 7 and we're looking for three consecutive numbers - having 5 & 7 already - then 6 is the next logical step since (6*5*7=210). Since the integers are 5, 6 and 7, their sum is 5 + 6 + 7 = 18.

11) $(x + 2)^2 + (x + 6)^2 = 2x^2$
$x^2 + 4x + 4 + x^2 + 12x + 36 = 4x^2$
Combine like terms:
$x^2 + x^2 + 4x + 12x + 4 + 36 - 4x^2 = 4x^2 - 4x^2$
Simplify: $-2x^2 + 16x + 40 = 0$
Divide both sides by -2: $x^2 - 8x - 20 = 0$
Factor $(x - 10)(x + 2)$
$x = 10$
The sides of the triangle are (10 + 2) (10 + 6) (2 x 10) or 12, 16, 20
Therefore, 12 + 16 + 20 = 48. The perimeter is 48 units.

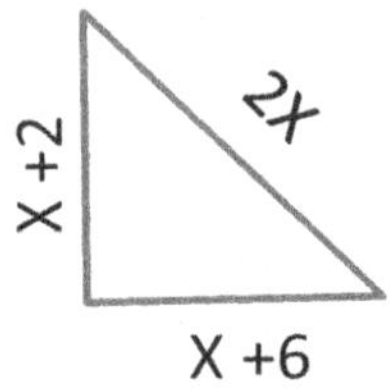

$a^2 + b^2 + c^2$

12) Use the order of operations PEMDAS. Solve the multiplication and division first.
6 + 2 x 7 – 8 ÷ 2 = 6 + (14) - (4) = 16

13) $x_1 = -2$, $y_1 = -1$, $x_2 = 4$, and $y_2 = 3$.
$\text{slope} = m = \frac{y_2 - y_1}{x_2 - x_1}$ = 3 – (-1)/ 4 – (-2) = 4/6 = 2/3

14) List the factors of both numbers and find the greatest.
1, 3, 7, 9, 21, <u>27</u>, 63, 189
1, 3, 9, 23, <u>27</u>, 69, 207, 621

15) 5.556×10^6 = 5,556,000 Move the decimal point to the right 6 places.

PERENNIAL MATH

INDIVIDUAL ANSWER KEY
(Grades 7-8)

Instructions for Scoring:

- ***Mark CORRECT answers by circling "O" the question number, and INCORRECT answers with an "X". Each correct answer is worth one point. Count the circles (correct answers) and write the number in the "SCORE" box.***

CORRECT → (1.) ______

INCORRECT → X. ______

1. **5/9**	6. **15**	11. **48**
2. **91**	7. **23 or 35**	12. **16**
3. **16**	8. **70**	13. **2/3**
4. **7**	9. **0**	14. **27**
5. **19**	10. **18**	15. **5,556,000**

Grade 7-8 Individual Event (30 Minutes)

1) A restaurant has a total of thirty-six 3-legged stools and 4-legged chairs. They have three times as many chairs as stools. What is the total number of legs on these stools and chairs?

2) A circle with radius 16 cm is inscribed in a square. The area of the shaded region is ____.

(Use π = 3.14 and round to the nearest whole number)

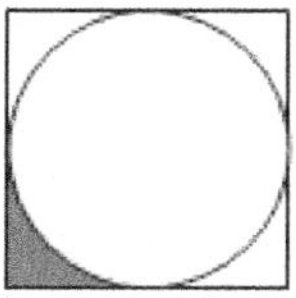

3) A circular spinner is divided into six equal sectors. The spinner arrow, initially pointing straight right, is spun counter-clockwise through 1999°. In which sector does the arrow stop?

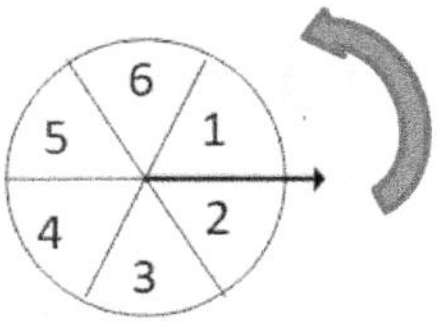

4) Willa buys donuts for \$6.60 per dozen. She spends \$16.50 for the donuts and then sells all of them for \$23.70. What is her average profit per donut?

5) In the sequence: 0, 1, 2, 5, 12, 29, ...; each number equals twice the previous number plus the number before that. For example, 29 = 2(12) + 5 . If 29 is the 6th number in this sequence, what is the 12th number in the sequence?

6) Rafa is eight years older than his sister. In three years, he will be twice as old as his sister. How old is Rafa now?

7) The fourteen members of the soccer team each have cell phones. If everyone has a text conversation going on with each of the other players at the same time, how many text conversations are going on. Only count conversation between player A and player B as 1 conversation. Do not count Player B with player A again.

8) In a school of 860 students, 440 are boys. One fourth of the students travel to school by bus. 300 of the girls do not ride the bus. How many of the boys ride the bus?

9) When Marvin applied for the job, the manager offered him \$0.01 on Day 1, \$0.02 on Day 2, \$0.04 on Day 3, \$0.08 on Day 4, and so on, doubling his pay each day. How much would Marvin earn on Day 20 alone?

10) Don completes 3/5 of a job in 45 minutes. Working at the same rate, how many more minutes will it take him to complete the job?

PERENNIAL MATH

11) Find the slope of a line passing through (2, 1) and (7, -2)

12) If x □ y is defined as $(x + y)^2 + 2xy$, then 3 □ 6 =

13) The prime factorization of 2000 can be written $2000 = 2^x 5^y$. The sum $x + y$ equals ____.

14) The marks on this number line are equally spaced. What is the value of N?

15) Convert 0.00000000578 to scientific notation.

INDIVIDUAL ANSWER DOCUMENT (Grades 7-8)

SCORE

Team # __________ School ______________________________

Student Name ______________________________________

Student Number D - ________________

1. ________	6. ________	11. ________
2. ________	7. ________	12. ________
3. ________	8. ________	13. ________
4. ________	9. ________	14. ________
5. ________	10. ________	15. ________

PERENNIAL MATH

Grade 7-8 Individual Event Solutions

1) A restaurant has a total of thirty-six 3-legged stools and 4-legged chairs. They have three times as many chairs as stools. To find the total number of legs on these chairs and stools, first solve for the number of chairs and stools they have. Use algebra and let x stand for the number of stools. Since there are a total of 36 and there are 3 times as many chairs as stools then there are $3x + x = 36$. $4x = 36$, so $x = 36/4$ or 9. If they have 9 stools and 27 chairs then the total number of legs is $9(3) + 27(4) = 27 + 108 = 135$ total legs.

2) A circle with radius 16 cm is inscribed in a square. To find the area of the shaded region subtract the area of the circle from the area of the square and then divide that by 4 because the shaded region is ¼ of that area. When a circle is inscribed in a square, the diameter of the circle is equal to the side length of the square. The length of the sides would be 32 cm. So the area of the square is 32^2 cm = 1024cm^2. The are of the circle is $\pi r^2 = 3.14 \times 16^2 = 803.84\ \text{cm}^2$. $1024\text{cm}^2 - 803.84\ \text{cm}^2 = 220.16\text{cm}^2$. Now divide by 4 to find the shaded region $220.16/4 = 55.04\ \text{cm}^2$ Round this to the nearest whole number = $55\ \text{cm}^2$

3) A circular spinner is divided into six equal sectors. So each sector is 60°. The spinner arrow, initially pointing straight right, is spun counter-clockwise through 1999° . 1999° divided by 60° = 33 with a remainder of 19°. So the spinner would pass through 33 complete revolutions and 19° into sector 4.

4) First divide \$16.50 by \$6.50 to see how many dozen she purchased. She purchase 2 and ½ dozen, which is 30 doughnuts. Now subtract the income from her original expense \$23.70 - \$16.50 = \$7.20. Now divide that total by 30 doughnuts. 7.20/30= .24 cents profit per doughnut.

5) In the sequence: 0, 1, 2, 5, 12, 29, ...; each number equals twice the previous number plus the number before that. Make a chart to find the 12th number in the sequence.

Number in the sequence	Actual Number	Formula 2x previous number + the number before the previous number
1	0	
2	1	
3	2	
4	5	
5	12	
6	29	
7	70	2(29) + 12
8	169	2(70) + 29
9	408	2(169) + 70
10	985	2(408) + 169
11	2378	2(985) + 408
12	5741	2(2378) + 985

6) Let x be a Rafa's age. Let x - 8 be his sister's age. In 3 years, Rafa's age will be x + 3. In 3 years, the sister's age will be x - 8 + 3 = x - 5
In 3 years, Rafa is twice as old as his sister.
$x + 3 = 2 \times (x - 5)$
$x + 3 = 2x - 10$
$2x - 10 = x + 3$
Subtract x from both sides.
$2x - x - 10 = x - x + 3$
$2x - x - 10 = 3$
Add 10 to both sides
$2x - x - 10 + 10 = 3 + 10$
$2x - x = 3 + 10$
$x = 13$
Rafa's age is 13 now.
His sister's age is 13 - 8 = 5.

In 3 years, Rafa's will be 16 and the sister will be 8.

16 is twice as old as 8.

PERENNIAL MATH

Grade 7-8 Individual Event Solutions

7) To calculate how many text conversations are going on between 14 players simultaneously, start with Player 1 and that player has a conversation going on with 13 people, now Player 2 has 12 other conversations going, Player 3 has 11 and so forth. So there are 13 + 12 + 11 + 10 + 9 + 8 + 7 + 6 + 5 + 4 + 3 + 2 + 1= 91 conversations.

Player 1	Player 2	Player 3	Player 4	Player 5	Player 6	Player 7	Player 8	Player 9	Player 10	Player 11	Player 12	Player 13
Player 2	Player 3	Player 4	Player 5	Player 6	Player 7	Player 8	Player 9	Player 10	Player 11	Player 12	Player 13	Player 14
Player 3	Player 4	Player 5	Player 6	Player 7	Player 8	Player 9	Player 10	Player 11	Player 12	Player 13	Player 14	
Player 4	Player 5	Player 6	Player 7	Player 8	Player 9	Player 10	Player 11	Player 12	Player 13	Player 14		
Player 5	Player 6	Player 7	Player 8	Player 9	Player 10	Player 11	Player 12	Player 13	Player 14			
Player 6	Player 7	Player 8	Player 9	Player 10	Player 11	Player 12	Player 13	Player 14				
Player 7	Player 8	Player 9	Player 10	Player 11	Player 12	Player 13	Player 14					
Player 8	Player 9	Player 10	Player 11	Player 12	Player 13	Player 14						
Player 9	Player 10	Player 11	Player 12	Player 13	Player 14							
Player 10	Player 11	Player 12	Player 13	Player 14								
Player 11	Player 12	Player 13	Player 14									
Player 12	Player 13	Player 14										
Player 13	Player 14											
Player 14												

8) In a school of 860 students, 440 are boys. One fourth of the students travel to school by bus. 300 of the girls do not ride the bus. To find out how many of the boys ride the bus make a chart to solve. First solve for how many girls there are. Next, solve for how many students travel by bus and how many do not travel by bus. Now subtract the number of girls who do not ride by bus and you will be able to see how many boys do not ride by bus. Now subtract the number of boys who do not ride by bus from the total number of boys and you will have 95 boys who ride the bus.

Total 860 students			
Boys	Girls	Bus = 1/4 total	No Bus = 3/4 total
440	860 - 440 = 420	860/4 = 215 students	860-215 = 645
		420 girls - 300 of the girls do not ride the bus = 120 girls who ride the bus	645 -300 girls who do not ride the bus = 345 boys who do not ride the bus
440 boys - 345 who do not ride the bus = 95 boys who ride the bus.			

PERENNIAL MATH

Grade 7-8 Individual Event Solutions

9) To find out how much Marvin made on Day 20 alone, make a chart and double the pay each day. He made $5242.88 on Day 20.

Day 1	0.01	Day 11	10.24
Day 2	0.02	Day 12	20.48
Day 3	0.04	Day 13	40.96
Day 4	0.08	Day 14	81.92
Day 5	0.16	Day 15	163.84
Day 6	0.32	Day 16	327.68
Day 7	0.64	Day 17	655.36
Day 8	1.28	Day 18	1310.72
Day 9	2.56	Day 19	2621.44
Day 10	5.12	Day 20	5242.88

10) Don completes 3/5 of a job in 45 minutes. If 3/5 = 45 min, then 1/5 = 15 min. The complete job would be 5/5. Add 2/5 or 30 more minutes to complete the job.

11) Find the slope of a line passing through (2, 1) and (7, -2) Use the slope intercept formula
$m = y_2 - y_1 / x_2 - x_1$
(-2-1)/7-2 = -3/5

12) If x □ y is defined as $(x + y)^2 + 2xy$, then
3 □ 6 = $(3 + 6)^2 + 2(3)(6)$ =
$9^2 + 36$ =
81 + 36 = 117

13) The prime factorization of 2000 can be written 2000 = $2^x 5^y$.
$2^4 5^3$ The sum x + y equals 4 + 3 = 7

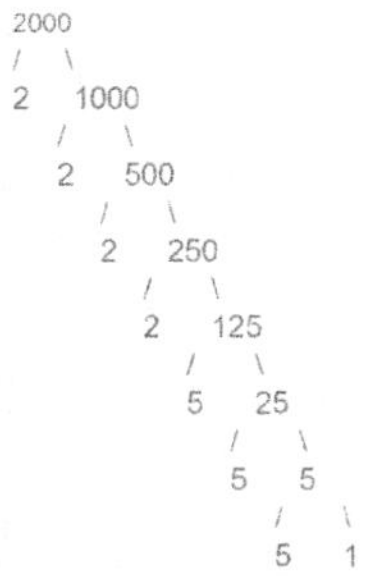

14) The marks on this number line are equally spaced. What is the value of N?

The distance from -230 to 190 = 420.
Divide this by 6 segments. 420/6 = 70.
190 – 2(70) = 190 - 140 = 50.

15) Convert 0.00000000578 to scientific notation. The decimal moves 9 places to the right. The answer is 5.78×10^{-9}

INDIVIDUAL ANSWER KEY
(Grades 7-8)

Instructions for Scoring:

- ***Mark CORRECT answers by circling "O" the question number, and INCORRECT answers with an "X". Each correct answer is worth one point. Count the circles (correct answers) and write the number in the "SCORE" box.***

CORRECT → (1.) ______

INCORRECT → X. ______

1. 135	6. 13	11. -3/5
2. 55	7. 91	12. 117
3. 4	8. 95	13. 7
4. .24 or 24	9. 5242.88	14. 50
5. 5741	10. 30	15. 5.78×10^{-9}

TEAM TESTS
GRADES 7-8

20 minute practice tests for teams
Each member gets a copy of the test, but the team only receives one answer document

Grade 7-8 Team Event (20 Minutes)

1) What is the sum of the first 7 prime numbers greater than 20?

2) The perimeter of one of the small rectangles in this shape is 24 cm and its length is twice the width. What is the area of the entire shape that is made up of these smaller rectangles?

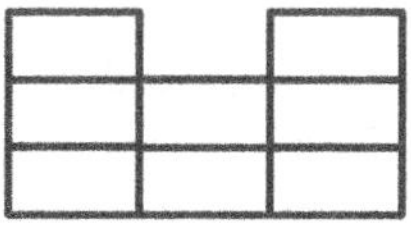

3) Find the surface area of this Rectangular Prism.

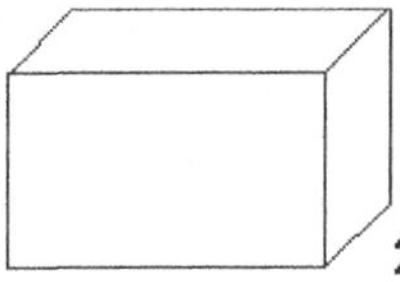

4) The average score for a class of 25 students was 75. If the 15 girls in the class averaged 70, then what was the average of the boys in the class?

5) What is the sum of all the odd composite numbers 1 through 25?

6) Trent can buy 2 pairs of pants and 1 shirt for $79.85 and 1 pair of pants and 2 shirts for $69.85, how much does one shirt cost?

7) In a tennis tournament there are 64 players in the draw. If this is a single elimination tournament, how many matches will the winner have to play in order to hold the championship trophy?

8) A collection of 33 coins, consisting of nickels, dimes, and quarters, has a value of $3.30. If there are three times as many nickels as quarters, and one-half as many dimes as nickels, how many nickels are there?

9) Find the slope of this graph.

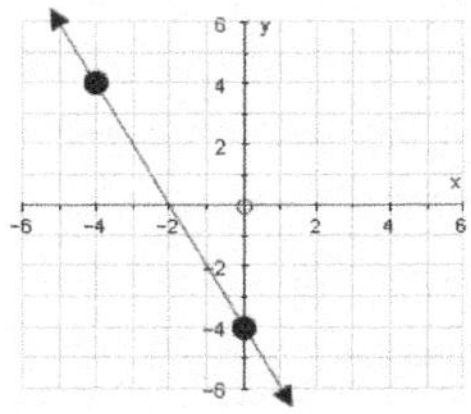

10) Ice Cream Dream sells sundaes with the choice of chocolate, butterscotch and nuts for toppings. One day they sold 256 total sundaes. Of these, 140 had butterscotch, 140 had chocolate, 84 had chocolate and nuts, 62 had butterscotch but no nuts, 68 had chocolate and butterscotch, 38 had all three toppings and 20 had no toppings. How many sundaes were sold with only nuts on them?

PERENNIAL MATH

TEAM ANSWER DOCUMENT
(Grades 7-8)

SCORE

Team # ___________ School ____

Student Names ____

____

Write answers clearly. Each correct answer is worth one point.

1. ____________

2. ____________

3. ____________

4. ____________

5. ____________

6. ____________

7. ____________

8. ____________

9. ____________

10. ____________

PERENNIAL MATH

Grade 7-8 Team Event Solutions

1) The sum of the first 7 prime numbers greater than 20 is (23 + 29 + 31 + 37 + 41 +43 + 47) = 251

2) The perimeter of one of the rectangles is 24cm and the length is twice the width, so the length must be 8cm and the width is 4cm. The area of one of the small rectangles would be 4cm x 8cm = 32cm^2. Therefore we can multiply the area of one of these rectangles by 8 to find the area of the entire shape is 32cm^2 x 8 = 256cm^2.

3) Find the area of two sides (Length x Height) x 2 sides
Find the area of adjacent sides (Width x Height) x 2 sides
Find the area of ends (Length x Width) x 2 ends
Add the three areas together to find the surface area
The surface area of a rectangular prism 7cm long, 5cm wide and 2cm high = (7 x 2 x 2) + (5 x 2 x 2) + (7 x 5 x 2) = 28 + 20 + 70 = 118 cm2.

4) The average score for the 25 students was 75, so the total score for all students had to be 25 x 75 = 1875. The average score for the 15 girls was 70, so their total score had to be 15 x 70 = 1050. The total score for the boys had to be 1875 - 1050 = 825. Since there were 10 boys, then their average score had to be 825 / 10 = 82.5.

5) The sum of all the odd composite numbers 1-25 is 9 +15 + 21 + 25 = 70

6) 2p + 1s = 79.85
1p + 2s = 69.85
3p+ 3s =149.70
Divide both sides by 3
p + s = 49.90

Subtract
1p + 2s = 69.85
- p + s = 49.90
s =19.95

7) Start with 64 players and make a chart or bracket: 64 ÷ 2 = 32; 32 ÷ 2 = 16; 16 ÷ 2 = 8; 8 ÷ 2 = 4; 4 ÷ 2 = 2; 2 ÷ 2 = 1. There are 6 games the winner must play to hold the trophy.

Game	Number of Players	Winners
1	64	32
2	32	16
3	16	8
4	8	4
5	4	2
6	2	1

PERENNIAL MATH

Grade 7-8 Team Event Solutions

8) Define your variables: number of quarters: q number of nickels: $3q$ number of dimes: $(½)(3q) = (3/2)q$

There is a total of 33 coins, so:

$q + 3q + (3/2)q = 33$

$4q + (3/2)q = 33$

$8q + 3q = 66$

$11q = 66$

$q = 6$

Then there are six quarters, and the number of nickels is 3q or 3(6) = 18 nickels.

9) Slope = $\frac{y2-y1}{x2-x1}$ (-4, 4) (0, -4)

$$\frac{(-4)-4}{0-(-4)} = \frac{-8}{4} = -2$$

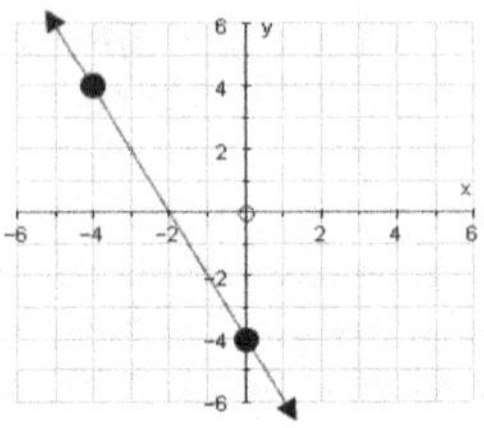

10) Use a Venn Diagram with 3 regions for chocolate, butterscotch, and nuts.

38 had all three toppings goes in the center. 84 had chocolate and nuts. The total number in the section that overlaps C and N should be 84. 38 is already in that circle, so 84 – 38 = 46 should be in the remaining piece. 68 had chocolate and butterscotch... The C and B overlap should total to 68...

So, 68 – 38 = 30 should go in the remaining. The 20 goes on the outside because it represents the sundaes with nothing. 140 had chocolate; 62 had butterscotch but no nuts. The C circle should total up to 140... So 140 – 30 – 38 – 46 = 26 is the number that should go in the remaining C spot. 62 should be the total in the B but no N section... There is already 30 in that section, so 62 – 30 = 32 should go in the B but no N section. 140 had butterscotch. In the remaining butterscotch slot should go 140 – 32 – 38 – 30 = 40. There are 256 total sundaes. So, we need to take 256 subtract all of the numbers to see how may are left for that last space....

256 – 20 – 40 – 46 – 38 – 30 – 32 – 26 = 24

The number that is in the N circle but no other circle: 24.

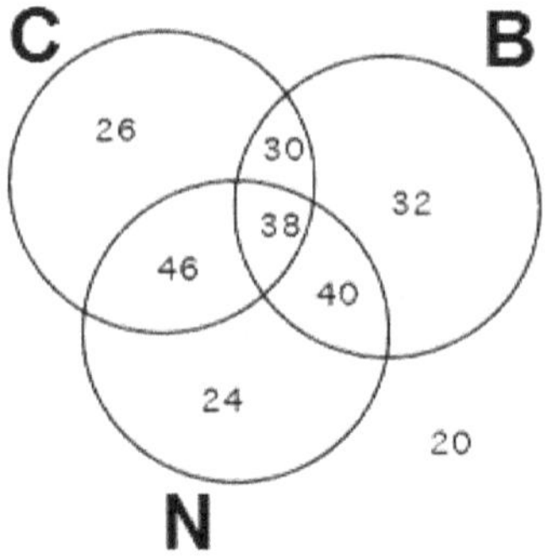

PERENNIAL MATH

TEAM ANSWER KEY
(Grades 7-8)

Instructions for Scoring:

- ***Mark CORRECT answers by circling "O" the question number, and INCORRECT answers with an "X" . Each correct answer is worth one point. Count the circles (correct answers) and write that number in the "SCORE" box.***

CORRECT → (1.) ________

INCORRECT → X. ________

1. **251**
2. **256 cm^2**
3. **118 cm^2**
4. **82.5**
5. **70**
6. **19.95**
7. **6**
8. **18**
9. **-2**
10. **24**

Grade 7-8 Team Event (20 Minutes)

1) In the dress shop the ratio of blue dresses to the number of red dresses is 7:13. What percent of the dresses are blue?

2) Find the measure in Angle A in the figure below

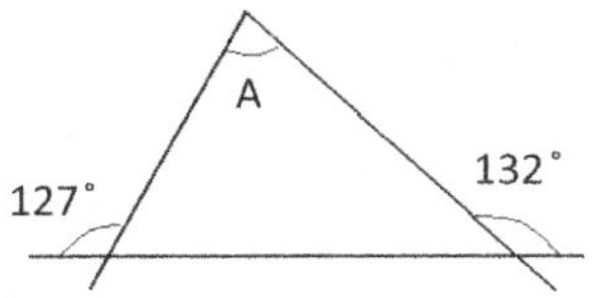

3) Find the area of the shaded region.

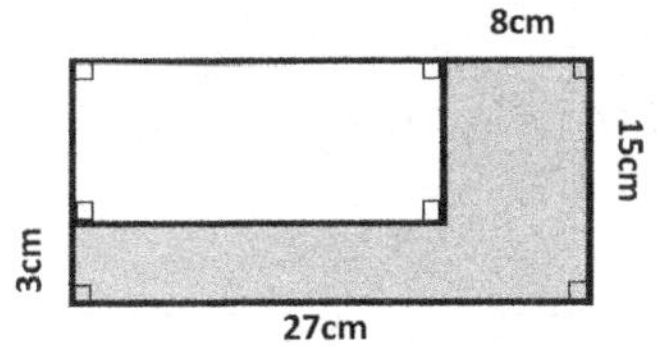

4) Meredith is twice as old as Jackson was three years ago. The total of their ages now is 36. How old is Jackson now?

5) What is the sum of all the prime numbers less than 100 that contain the digit 7?

6) Trent can buy 2 pencils and 1 eraser for 97 cents and 1 pencil and 2 erasers for 83 cents, how much does one eraser cost?

7) In a tennis tournament there are 128 players in the draw. If this is a single elimination tournament, how many matches will the winner have to play in order to hold the championship trophy?

8) Clyde counted the flamingos and giraffes at the zoo. He noticed there were three and a half times the number of legs as the number of heads and the total of the legs and heads was 72. How many giraffes were counted?

9) A Class of 50 students completed a survey on what pets they have. Everyone had at least one pet. 4 students had Cats and Fish but not Dogs. 12 students had Dogs and Fish but not Cats. 10 students had Cats and Fish. 18 students had Cats and Dogs. Altogether, 30 students had Cats, 32 students had Dogs, and 28 students had Fish. How many students had fish only?

10) Using any combination of half dollars, quarters, dimes, and nickels how many ways can you make 85 cents?

TEAM ANSWER DOCUMENT
(Grades 7-8)

SCORE

Team # ___________ School ______________________________________

Student Names____

____

Write answers clearly. Each correct answer is worth one point.

1. ____________
2. ____________
3. ____________
4. ____________
5. ____________
6. ____________
7. ____________
8. ____________
9. ____________
10. ____________

PERENNIAL MATH

Grade 7-8 Team Event Solutions

1) The ratio of blue dresses to the number of red dresses is 7:13. Assume there are 20 total dresses, so 7 are blue and 13 are red. Therefore, the percent that of blue dresses is 7/20 or 35/100 which is 35%.

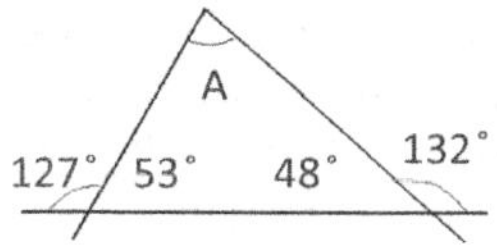

2) Subtract to find the values of the inside angles first. $180^\circ - 127^\circ = 53^\circ$ and $180^\circ - 132^\circ = 48^\circ$. Now add those angles and subtract from 180°.
$180^\circ - (53^\circ + 48^\circ) = 79^\circ$

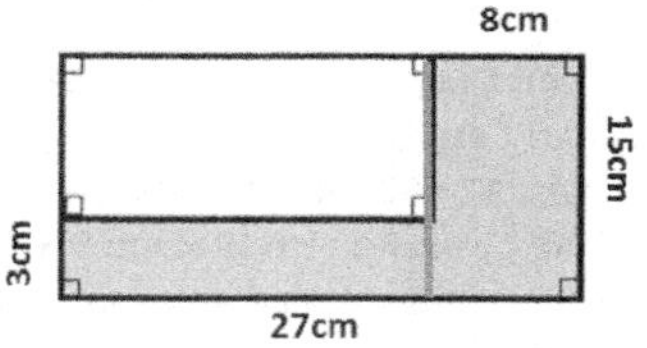

3) Break the shape up into smaller shapes and find the areas of the smaller shapes and add them up. One of the areas is 8cm x 15cm = $120cm^2$. The next area is (27 -8 = 19cm) x 3cm. 19cm x 3cm = $57cm^2$. Now add those smaller areas together: $120cm^2$ + $57cm^2$ = $177cm^2$

4) Meredith is twice as old as Jackson was three years ago. The total of their ages is 36.
Meredith = 2 (J -3) so

$2(J-3) + J = 36$
$2J - 6 + J = 36$
$3J - 6 = 36$
$3J - 6 + 6 = 36 + 6$
$3J = 42$
$J = 42/3$ OR 14

Jackson is 14 now. Check answer by calculating Meredith's age now. Remember that Meredith is 2(J-3) or 2(11) = 22 years old, which is twice as old as Jackson's age 3 years ago.

5) The sum of all the primes under 100 with a digit 7 in them is :

$$7 + 17 + 37 + 47 + 67 + 71 + 73 + 79 + 97 = 495$$

6) $2p + 1e = .97$
$-\ \underline{1p + 2e = .83}$
$3p + 3e = 1.80$
Divide both sides by 3
$p + e = .60$

Subtract this from the know equations above to determine the value of the eraser
$1p + 2e = .83$
$\underline{p + e = .60}$
$e = .23$

7) Start with 128 players and divide by 2 for each match played until you get to one, the champion:
128 ÷ 2 = 64; 64 ÷ 2 = 32; 32 ÷ 2 = 16; 16 ÷ 2 = 8; 8 ÷ 2 = 4; 4 ÷ 2 = 2; 2 ÷ 2 = 1.
There are 7 games that the winner must play.

Grade 7-8 Team Event Solutions

8) Use algebra. There are 3 ½ times as many legs as heads. Let x represent heads. So the total legs and heads can be represented with the equation: $3.5x + x = 72$, $4.5x = 72$, $x = 72/4.5 = 16$ heads. Working from 16 heads, you can make a chart to calculate legs. Start with 8 of each animal and notice that the total heads and legs is less than 72, so increase the number of giraffes (decrease the number of flamingos) each time until you reach 72. There are 12 giraffes and 4 flamingos.

Flamigos	8	6	4
Giraffes	8	10	12
Heads (H)	16	16	16
Legs (L)	48	52	56
Total (H+L)	64	68	72

9) Use a Venn Diagram. All regions total 50 students. Label the totals for each category:30 students had Cats, 32 had Dogs, and 28 had Fish. Fill in these areas: 4 students had Cats and Fish but not Dogs. 12 students had Dogs and Fish but not Cats. Use these to solve for missing areas: 10 students had Cats and Fish [Region V + 4]. 18 students had Cats and Dogs [Region V + 12]. Once you solve for Region V then you can solve for the fish to get a total of 28. There are 6 students who had fish only.

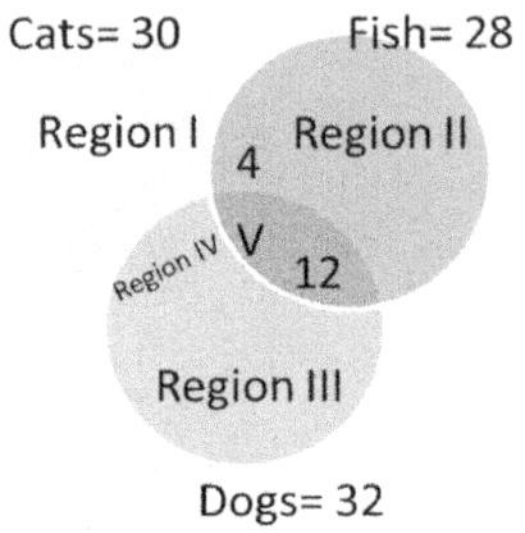

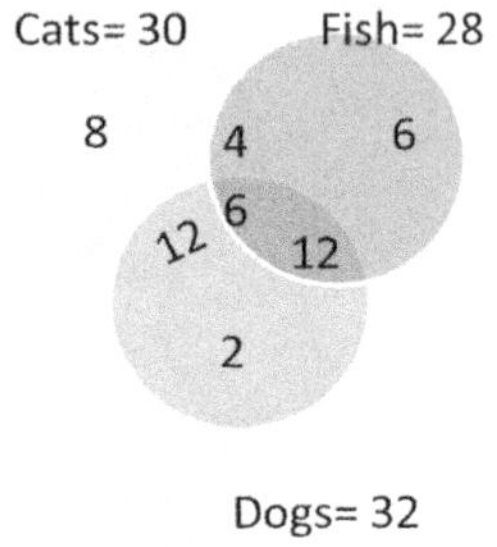

10) Make an organized list. There are 28 combinations.

Half Dollar	Quarters	Dimes	Nickels
1	1	1	0
1	1	0	2
1	0	3	1
1	0	2	3
1	0	1	5
1	0	0	7
0	3	1	0
0	3	0	2
0	2	3	1
0	2	2	3
0	2	1	5
0	2	0	7
0	1	6	0
0	1	5	2
0	1	4	4
0	1	3	6
0	1	2	8
0	1	1	10
0	1	0	12
0	0	8	1
0	0	7	3
0	0	6	5
0	0	5	7
0	0	4	9
0	0	3	11
0	0	2	13
0	0	1	15
0	0	0	17

PERENNIAL MATH

TEAM ANSWER KEY
(Grades 7-8)

Instructions for Scoring:

- ***Mark CORRECT answers by circling "O" the question number, and INCORRECT answers with an "X". Each correct answer is worth one point. Count the circles (correct answers) and write that number in the "SCORE" box.***

CORRECT → (1.) ____________

INCORRECT → X. ____________

1. **35**
2. **79**
3. **177**
4. **14**
5. **495**
6. **.23**
7. **7**
8. **12**
9. **6**
10. **28**

Grade 7-8 Team Event (20 Minutes)

1) The average of 13 and x is equal to the average of 7, 19, and x. What is the value of x?

2) In this magic square all rows, columns, and diagonals have the same sum. What value should X have?

4	11		
	13	12	X
9	2	7	
		1	10

3) The area of this shape is 252cm². What is the perimeter of the shape?

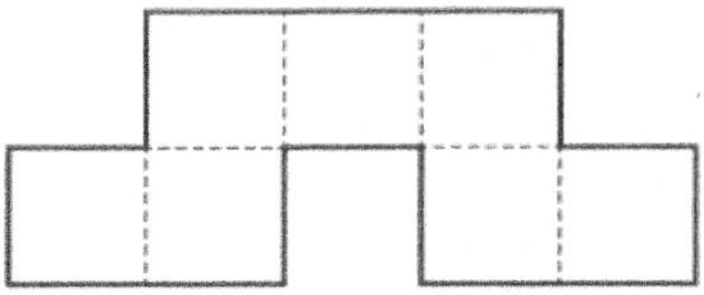

4) The average age of the 6 people in room A is 40. The average age of the 4 people in room B is 25. If the people in the two rooms gather together, what is the average age of the 10 people?

5) Last year 150 students attended the Summer Camp. Of those 150 campers, 96 were girls and 54 were boys. Also, 68 were 7th graders and 82 were 8th graders. How many 8th-grade boys attended the camp if 40 of the girls were 7th graders?

6) In a tennis tournament there are 64 players in the draw. If this is a single elimination tournament, how many matches will the winner have to play in order to hold the championship trophy?

7) The pet shop has 100 finches to put in cages. Each cage must contain at least one finch and no cage can contain the same number of finches. What is the maximum number of cages that could be required?

8) If Anthony gets a 97 on his next math test, his average will be 89. However, if he gets 73, his average will be 85. How many tests has Anthony already taken?

9) The sum of 9 consecutive whole numbers is 270. What is the product of the first three of these consecutive whole numbers?

10) Mick throws four darts at a target with scoring regions of 3, 4, and 5 points. If all four darts hit a scoring region, how many different point totals are possible?

 PERENNIAL MATH

TEAM ANSWER DOCUMENT
(Grades 7-8)

SCORE

Team # __________ School ____________________________________

Student Names ____

____

Write answers clearly. Each correct answer is worth one point.

1. ____________
2. ____________
3. ____________
4. ____________
5. ____________
6. ____________
7. ____________
8. ____________
9. ____________
10. ____________

PERENNIAL MATH

Grade 7-8 Team Event Solutions

1) The average of 13 and x is equal to the average of 7, 19, and x. So (x +13)/2 = (7 + 19 +x)/3. Cross multiply. 52+2x=3x+39. Therefore, 13 = x

2) Look for shared squares. 11 + 13 + 2 = __ + 1 + 10. 15 goes in the bottom left missing square. Next you can solve for X because the first square in the second row is a shared square. 4 + 9 + 15 = 13 + 12 + X. The number 3 is what should go in the blank for X.

4	11		
	13	12	X
9	2	7	
15		1	10

4	11		
	13	12	3
9	2	7	
15		1	10

3) Since the area of the shape is 252cm², divide by 7 to find the area of one of the squares. Each square has an area of 36cm². Therefore the length of each side of the small squares is 6cm. There are 16 sides going around the perimeter of the shape. 16 x 6cm = 96cm.

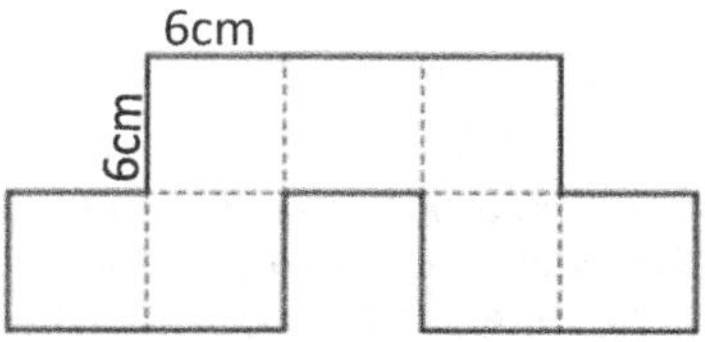

4) The average age of the 6 people in room A is 40. So the sum of their ages is 240. The average age of the 4 people in room B is 25. The sum of their ages is 100. 240 + 100 = 340. 340/10 = 34. The average age of the 10 people in the room is 34.

5) Of the 150 campers, 96 were girls and 54 were boys. Also, 68 were 7th graders and 82 were 8th graders. Since 40 of the girls were 7th graders, subtract 40 from 96 to find out how many girls were 8th graders. 56 girls were 8th graders. Since there were 82 8th graders, subtract the 8th grade girls 82-56 = 26 boys in the 8th grade.

6) Start with 64 players and divide by 2 for each match played until you get to one, the champion: 64 ÷ 2 = 32; 32 ÷ 2 = 16; 16 ÷ 2 = 8; 8 ÷ 2 = 4; 4 ÷ 2 = 2; 2 ÷ 2 = 1.
There are 6 games that the winner must play.

7) If you start with the first cage and work your way up 1 + 2 + 3...+ 13 you will have 91 finches with 9 finches left over. You could equally distribute those 9 finches or put them in the 13th cage to make it contain 22 finches. There are not enough finches to fit in a 14th cage without having an identical number of finches.

Grade 7-8 Team Event Solutions

8) If Anthony gets a 97 on his next math test, his average will be 89. However, if he gets 73, his average will be 85. How many tests has Anthony already taken? To get an average of 89, you can take the sum of all tests and divide by the number of tests. So using guess and check, make a chart. If he had taken 6 tests and had an average of 89 his sum would be 534. Subtract the 97 he made on the 6th test and he had 437 points to begin with. If you added 73 to that he would have 510 and 510/6 = an 85 average. So Anthony had already taken 5 tests.

Number of Tests	Average	Total Sum	Original Sum of tests (-97)	Add 73	Average
1	**89**	89	-8	65	65
2	**89**	178	81	154	77
3	**89**	267	170	243	81
4	**89**	356	259	332	83
5	**89**	445	348	421	84.2
6	**89**	534	437	510	**85**

9) The sum of 9 consecutive whole numbers is 270. What is the product of the first three of these consecutive whole numbers? 270 ÷ 9 = 30. The mean is 30 so the 9 consecutive numbers are 26, 27, 28, 29, 30, 31, 32, 33, 34. The product of the first 3 numbers is 26 x 27 x 28 = 19,656.

10) Make an organized list to all the point values and to account for repetition. There are 9 different possible point values.

Dart 1	Dart 2	Dart 3	Dart 4	Points
3	3	3	3	12
3	3	3	4	13
3	3	3	5	14
3	3	4	4	14
3	3	4	5	15
3	3	5	5	16
3	4	4	4	15
3	4	4	5	16
3	4	5	5	17
3	5	5	5	18
4	4	4	4	16
4	4	4	5	17
4	4	5	5	18
4	5	5	5	19
5	5	5	5	20

TEAM ANSWER KEY
(Grades 7-8)

__Instructions for Scoring:__

- ***Mark CORRECT answers by circling "O" the question number, and INCORRECT answers with an "X". Each correct answer is worth one point. Count the circles (correct answers) and write that number in the "SCORE" box.***

CORRECT → (1.) ________

INCORRECT → X. ________

1. 13
2. 3
3. 96
4. 34
5. 26
6. 6
7. 13
8. 5
9. 19,656
10. 9

Grade 7-8 Team Event (20 Minutes)

1) What is the sum of the first 10 prime numbers greater than 20?

2) In this square, the radius of each half circle is 6 inches. What is the area of the shaded region? Use 3.14 for π.

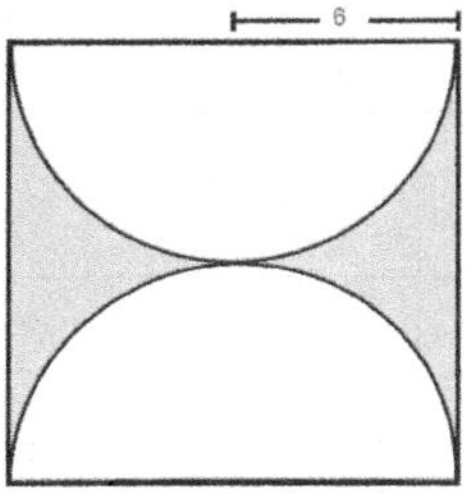

3) A 26 ft ladder is leaned up against a wall. It is 10 ft from the base of the wall. How far up does the ladder reach?

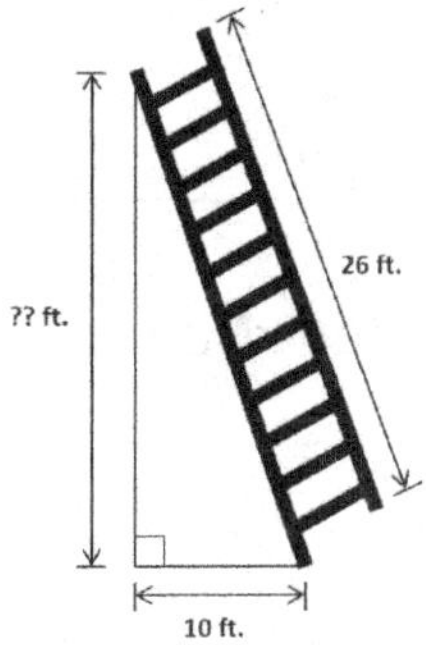

4) What is the probability for a couple to have 5 children in a row of the same gender? Express answer in simplified fraction form.

5) Nine pennies weigh one ounce. Mary has 5 pounds of pennies and Brett has 3.5 pounds of pennies. How much more money in dollars and cents does Mary have than Brett?

6) How many two-digit and three-digit positive integers less than 160 have the units digit greater than the tens digit?

7) Carl jogged around a one-fourth mile track at 5 mph. If he jogged for 42 minutes, how many times did he go around the track?

8) In a triangle, one angle is three times as large as the other and the third is 100° greater than the sum of the other two. How many degrees are in the smallest angle of the triangle?

9) The sum of 4 consecutive odd integers is 136. What is the product of these four consecutive integers?

10) At a party, one-third of the guests were men, one-fourth were boys, one sixth were women, and 6 were girls. How many guests were at the party?

PERENNIAL MATH

TEAM ANSWER DOCUMENT
(Grades 7-8)

SCORE

Team # __________ School ______________________________________

Student Names ____

____

Write answers clearly. Each correct answer is worth one point.

1. ____________

2. ____________

3. ____________

4. ____________

5. ____________

6. ____________

7. ____________

8. ____________

9. ____________

10. ____________

2017-18-2

Grade 7-8 Team Event Solutions

1) The sum of the first 10 prime numbers greater than 20 are:

$$23 + 29 + 31 + 37 + 41 + 43 + 47 + 53 + 59 + 61 = 424$$

2) Subtract the area of the circle from the area of the square to find area of the shaded region.

The area of the square is $12^2 = 144$

The area of the circle is $\pi r^2 = 3.14 \times 6^2 = 113.04$

The area of the shaded region is $144 - 113.04 = 30.96 \text{ in}^2$

3) Use the Pythagoras Theorem to solve.

$$a^2 + b^2 = c^2$$
$$10^2 + b^2 = 26^2$$
$$100 + b^2 = 676$$
$$b^2 = 576$$
$$b = 24$$

The ladder reaches up 24 feet.

4) Use two rules. First use the rule of multiplication:

B B B B B = ½ x ½ x ½ x ½ x ½ = 1/32

G G G G G = ½ x ½ x ½ x ½ x ½ = 1/32

Now use the rule of addition:

1/32 + 1/32 = 2/32 = 1/16

5) To find how much more money in dollars and cents that Mary has than Brett, first convert the pounds of pennies to ounces and then multiply by 9 to see how many pennies each had.

Mary has 5 lbs x 16 = 80 oz. of pennies. 80 x 9 = 720 pennies

Brett has 3.5 lbs. x 16 = 56 oz. of pennies. 56 x 9 = 504 pennies

Now determine the difference between Mary and Brett.

720 pennies – 504 pennies = 216 pennies.

Convert to dollars and cents, so Mary had $2.16 more than Brett.

6) Make a chart to see the possible numbers: there are 75 numbers that have the units digit greater than the tens digit.

159	149	139	129	119	109	89	79	69	59	49	39	29	19	
158	148	138	128	118	108		78	68	58	48	38	28	18	
157	147	137	127	117	107			67	57	47	37	27	17	
156	146	136	126	116	106				56	46	36	26	16	
	145	135	125	115	105					45	35	25	15	
		134	124	114	104						34	24	14	
			123	113	103							23	13	
				112	102								12	
					101									
4	5	6	7	8	9	1	2	3	4	5	6	7	8	75

PERENNIAL MATH

Grade 7-8 Team Event Solutions

7) Carl jogged around a one-fourth mile track at 5 mph. If he jogged for 42 minutes, how many times did he go around the track?
First find out how far Carl can jog in 42 minutes using ratios:

$$\frac{5\text{ miles}}{1\text{ hours}} = \frac{5\text{ miles}}{60\text{ minutes}} = \frac{x\text{ miles}}{42\text{minutes}} \qquad \textit{cross multiply}$$

$60x = (5)(42)$
$x = 210/60$ or
$x = 3.5$ miles in 42 minutes

Next determine the number of laps with 3.5 miles / 0.25 miles = 14 laps

8) In a triangle, one angle is three times as large as the other and the third is 100° greater than the sum of the other two. How many degrees are in the smallest angle of the triangle? Let x be the first angle. Then the second angle is three times the first angle or 3x. The third angle is then (x + 3x) + 100. Since all three angles must add up to 180 degrees we have the following equation:

$x + 3x + (x + 3x) + 100 = 180$ solve for x
$8x + 100 = 180$
$x = 10$

Therefore the smallest angle is the first angle, x, which is 10 degrees.

9) Use algebra to find the four consecutive odd integers. Let x represent the first integer, then the other three integers are x + 2, x + 4, and x + 6. Therefore

$x + x + 2 + x + 4 + x + 6 = 136$
$4x + 12 = 136$
$4x = 124$
$x = 31$

The four consecutive integers are 31, 33, 35, and 37. Their product is 31 x 33 x 35 x 37 = 1,324,785

10) At a party, one-third of the guests were men, one-fourth were boys, one sixth were women, and 6 were girls. How many guests were at the party?
Solve using proportions. First determine the girls' proportion of the guests.

$1/3 + 1/4 + 1/6 + g = 1$ *using 24 as the LCD*
$8/24 + 6/24 + 4/24 + g = 1$ *solve for g*
$g = 6/24$

Now determine the total guests using the LCD 24 as the total proportions. Since there are 6 girls and they make up $6/24^{th}$ of the guests, then each proportion is one (1) guest. Therefore, there are 8 men, 6 women, 4 boys, and 6 girls for a total of 24 guests at the party.

TEAM ANSWER KEY
(Grades 7-8)

Instructions for Scoring:

- ***Mark CORRECT answers by circling "O" the question number, and INCORRECT answers with an "X" . Each correct answer is worth one point. Count the circles (correct answers) and write that number in the "SCORE" box.***

CORRECT → (1.) ______

INCORRECT → X. ______

1. **424**
2. **30.96**
3. **24**
4. **1/16**
5. **2.16**
6. **75**
7. **14**
8. **10**
9. **1,324,785**
10. **24**

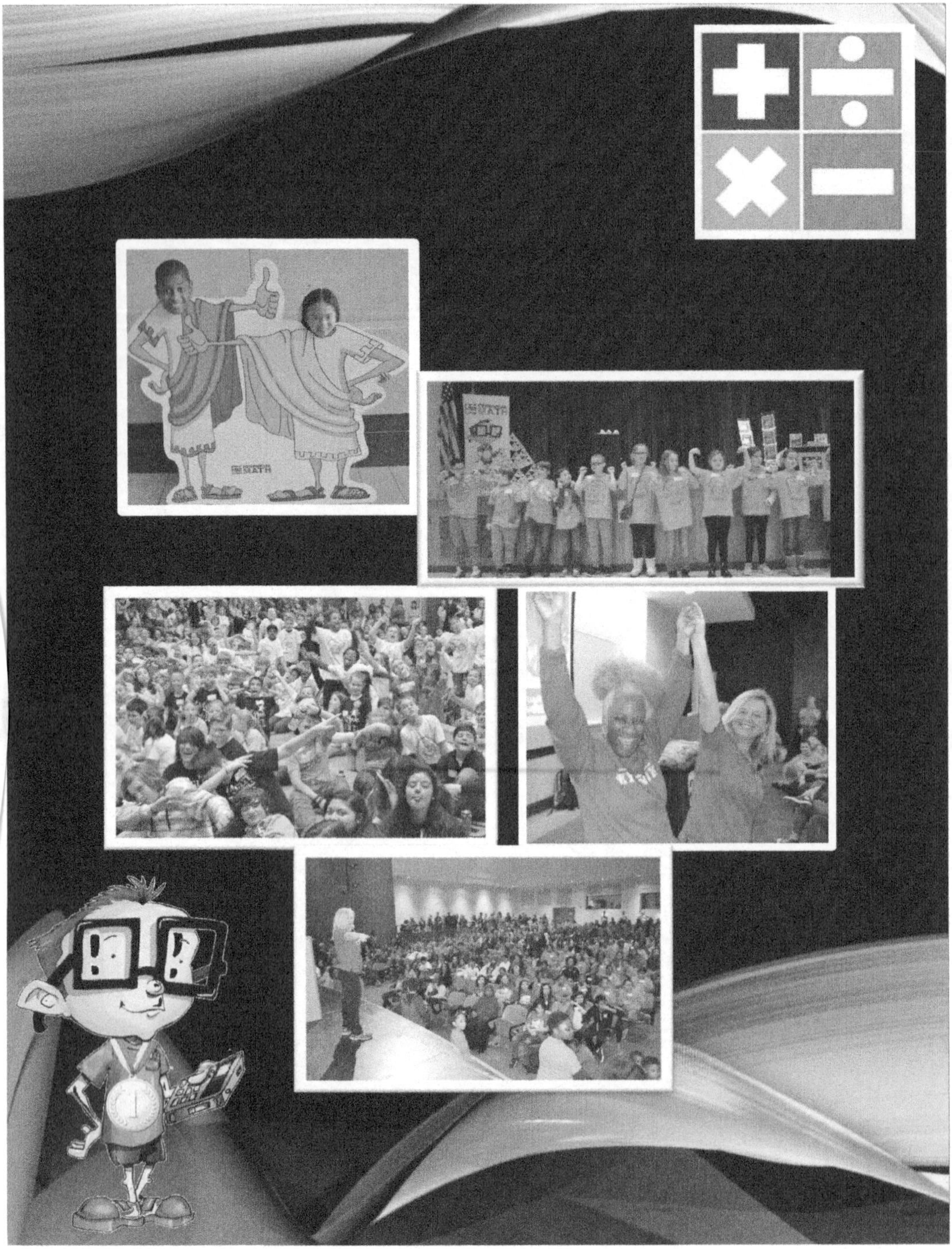

Made in the USA
Monee, IL
30 March 2026

47282686R00036